Josiah Keep

Botanical record book

containing directions for laboratory work in botany

Josiah Keep

Botanical record book
containing directions for laboratory work in botany

ISBN/EAN: 9783337711948

Printed in Europe, USA, Canada, Australia, Japan

Cover: Foto ©berggeist007 / pixelio.de

More available books at **www.hansebooks.com**

THE

BOTANICAL RECORD BOOK

CONTAINING

DIRECTIONS FOR LABORATORY WORK

IN BOTANY, LIST OF BOTANICAL TERMS, SPACES FOR DRAWINGS AND OBSERVATIONS, PREPARED BLANKS FOR RECORDING THE ANALYSES OF PLANTS, ETC.

PREPARED FOR THE USE OF SCHOOLS

BY

JOSIAH KEEP, A. M.,

Professor of Natural Science,
MILLS COLLEGE.

Published by the Author, Mills College, Alameda County, California.

SAN FRANCISCO:
H. S. CROCKER & CO., STATIONERS AND PRINTERS,
215, 217 and 219 Bush Street.

1890.

INTRODUCTION.

Whatever can aid either teacher or pupil in the study of Botany is worthy of attention. Especially is this true when the promised aid lies in the direction of making simple a series of actual experiments and observations upon plants and vegetable productions. There is very much to learn in the short time usually allotted to the study of this science,—the structure and morphology of the different parts and organs of the plant, the meaning of many new terms, the methods of analysis and classification, and the main features of the chief botanical orders or families. Moreover, the study of books and charts is not sufficient; but the pupil should see and handle the objects of which the lesson treats, and, if possible, should gather living specimens on the hills and in the meadows. The characteristics of the plant having been studied, a specimen should be carefully pressed, and at length should find a permanent resting place in that botanical casket, the student's herbarium. All of this work takes time; hence the course should be laid out orderly and with care.

The following pages have been prepared with a view to attaining the most complete results consistent with the amount of time which the student can ordinarily devote to this study. If a record page is crowded with spaces requiring detailed entries, the pupil is apt to be discouraged by the formidable task, and to slight or wholly omit certain portions; besides, the writing must be cramped on account of insufficient space. In the following blanks enough space has been provided so that the penmanship may be clear; and, while the essential features of description are retained, the list is not made so exhaustive as to appear discouraging.

It is believed by the author, after a considerable experience in teaching this science, that for most of our pupils such a course is to be preferred to one which demands numerous and obscure details.

In the preparation of the topics and blanks for laboratory work, the same rule has been kept in mind; the attention has been directed to the most important features, and convenient spaces have been prepared for the records.

Suggestions Concerning Laboratory Work.

The work indicated on the following pages is of great value in the study of Botany, and none of the ten subjects should be omitted. If the available time is too limited for them all, a less number of specimens than is recommended for each subject may be examined; on the other hand, additional work can be devised by the teacher who has an abundance of time at his command.

The various subjects should be taken up soon after they have been studied in the text-book, while the interest of the pupil is keen. The collection of flowers for the herbarium should be begun as early in the season as good specimens can be obtained; and observations upon the same should be recorded so far as the pupils understand the terms. Any blank spaces can usually be filled later in the season. It is hardly necessary to say that nothing should be recorded which has not been actually seen by the pupil who is making the record. It is desirable that the teacher should give some simple instruction in drawing. Outline drawings at least can be made by all, while the shading may be added by those who are able to do so. Every pupil should make the drawings, endeavoring to express what has been really seen. While all should try to draw well, the teacher will make allowance for differences in artistic taste and skill. If this plan is faithfully carried out, the Record Book may become an object of much interest, and perhaps of beauty also.

In the study of solid objects, like seeds, buds, etc., three drawings should be made:—the first, of the object as a whole; the second, of a longitudinal section along the axis; while the third should represent a transverse section through the center. In the case of a bud, *e. g.*, the first will represent it on the twig, the second will clearly show the relation of the scales and immature leaves to the axis, and the third will present their relation to one another. Good earnest work in this part of the study of Botany will amply repay the patient student; and the teacher who leads the pupil on in the investigation of these interesting phases of the Creator's handiwork will not fail of a reward.

The tools needed by each pupil are a lens, or simple microscope, a sharp knife, a few needles set in handles, and a small metric rule for taking measurements. In the schoolroom there should be a compound microscope, a hone for sharpening knives and needles, and a few boxes for holding material and apparatus. Of the many presses used for preparing plants for the herbarium, the author has found none of such general use as the lattice-work press made by Jas. W. Queen & Co., of Philadelphia.

Schedule of Exercises.

In each case, the student will follow the outline for the subject as given on the following pages.

1. Plant several kinds of seeds, and record observations.
2. Study six seeds according to the directions given.
3. Examine four buds; make three drawings of each.
4. Study a root, a bulb and a tuber.
5. Study ten leaves; make one drawing of each.
6. Observe four fruits, making proper records and drawings.
7. Make and study transverse sections of the stem of an exogen and of an endogen.
8. Examine with the compound microscope the pollen of several flowers; make drawings, and record shape, color, etc.
9. Record in this book the required observations on forty different species of plants.
10. Make an herbarium of at least fifty specimens of dried and mounted plants. Put on each a number corresponding with its page in the Record Book.

OUTLINES FOR LABORATORY WORK.

I. PLANT RAISING.

At the beginning of the term let each pupil plant a few seeds, either in a garden or in a flower pot, and then watch the growth of the seedlings, draw the plantlets, and record observations. The seeds of flax and of the Morning Glory are recommended, also beans, peas, and squash seeds.

NOTE.—On the blank pages which follow, the column for drawings is always numbered (1), while the lines are numbered from 2 to 10. This is done for uniformity, and does not mean that the drawings should always be made first. The record for each topic, however, should be placed on the line whose number corresponds with that of the topic. In this way explanatory writing will be avoided, as each answer can be referred to its proper question or topic. The record or answer should usually be brief and concise, and should express in a few words the main points of the object under consideration.

TOPICS.

1. Drawings of plantlets.
2. Date of planting.
3. Name and number of seeds.
4. Dates when plantlets came up.
5. Their manner of growth.
6. Details of culture.
7. Date of flowering.
8–10. Results, for each kind.

II. SEEDS.

Use large, well-formed seeds; prepare by soaking in water for twelve hours or more. Select such a variety of seeds as will best exhibit the chief forms of embryo and albumen. The following are recommended and in the order mentioned: Squash seeds, beans, peas, buckeyes or horse chestnuts, Morning Glory seeds, corn, and pine nuts.

TOPICS.

1. Drawings.
2. Name of seed.
3. Size (in millimeters).
4. Surface.
5. Hilum, shape and position.
6. Coats.
7. Cotyledons.
8. Caulicle.
9. Plumule.
10. Albumen.

III. BUDS.

Select a few large, well-formed buds, like those of the buckeye, walnut, cherry, maple and willow. Make a drawing of the bud on the twig, also of a longitudinal and of a transverse section.

TOPICS.

1. Drawings.
2. Name.
3. Size.
4. Arrangement on twig.
5. Color.
6. Varnish, etc.
7. Scales.
8. Interior structure.
9. Leaf bud or flower bud?
10. Remarks.

IV. ROOTS, TUBERS AND BULBS.

A carrot, a potato, and an onion will serve as typical specimens. The structure and peculiarities of each should be carefully noted. Make drawings of each as a whole, also of sections, both transverse and longitudinal.

TOPICS.

1. Drawings.
2. Name and kind.
3. Size.
4. Color { External. Internal.
5. Surface.
6. Texture.
7. Internal structure.
8. Bud or buds.
9. Rootlets.
10. Remarks.

V. LEAVES.

Select various kinds of leaves to illustrate the different forms. If the leaf is compound, describe one leaflet, and show the form of the whole by the drawings.

TOPICS.

1. Draw outline and venation.
2. Name.
3. Size.
4. Surfaces.
5. Shape.
6. Margin.
7. Apex and base.
8. Venation.
9. Petiole and stipules.
10. Microscopic structure.

VI. FRUITS.

Take such as may be obtained easily; for example, apples, cranberries, pods of the pea, radish, or mustard, mallow "cheeses," etc. Remember that "The ovary matures into the fruit."

TOPICS.

1. Drawings.
2. Name and class.
3. Size.
4. Color.
5. Pericarp.
6. Number of seeds.
7. Their position.
8. Stem.
9. What part of this fruit is of value?
10. Remarks.

VII. STEMS.

Sections of stems, both herbaceous and woody, can easily be made with a knife or saw, and are very interesting, especially when examined with a lens or a compound microscope.

TOPICS.

1. Drawings.
2. Name of plant.
3. Kind of stem.
4. Size.
5. Age.
6. Pith.
7. Wood, how arranged?
8. Bark.
9. Medullary rays.
10. Ducts.

VIII. POLLEN.

The study of pollen will depend largely upon the microscope at command; but many interesting facts may be learned by observation with a simple lens.

TOPICS.

1. Drawings.
2. From what flower?
3. Size.
4. Color.
5. Abundance.
6. Are the flowers perfect?
7. Are all the pollen grains alike?
8. How distributed?
9. Adhesion to stigma.
10. Remarks.

Subjects IX and X are provided for in the latter part of this book. They should be taken up as early in the term as circumstances permit; and their faithful study will bring much pleasure to both pupil and teacher.

TERMS MOST COMMONLY USED IN THE DESCRIPTION OF PLANTS.

STEM.

Class,—exogenous, endogenous.

Character, — herbaceous, suffrutescent, suffruticose, fruticose, arborescent, arboreous.

LEAF.

Insertion,—alternate, opposite, whorled.

Venation,—pinni-netted, palmi-netted, parallel-veined.

Form,—linear, lanceolate, falcate, oblong, elliptical, oval, ovate, orbicular, oblanceolate, spatulate, deltoid, cuneate, cordate, obcordate, reniform, sagittate, hastate, peltate, pinnately compound, palmately compound.

Margin,—Entire, serrate, dentate, crenate, wavy, incised, lobed, cleft, parted, divided.

Surface,—smooth, glabrous, pubescent, hirsute, hispid.

FLOWER.

Inflorescence,—solitary, raceme, corymb, umbel, spike, head, spadix, catkin, panicle, cyme, fascicle, glomerule, scorpioid.

Perfectness,—perfect, monœcious, diœcious.

Regularity,—regular, irregular.

Cohesion,—polypetalous, gamopetalous, apetalous.

Shape of corolla,—rosaceous, cruciform, papilionaceous, anomalous, rotate, campanulate, funnel-shaped, tubular, salver-shaped, labiate, ligulate.

Insertion of stamens,—hypogynous, perigynous, epigynous, epipetalous, gynandrous.

OVARY.

Cohesion of carpels,—distinct, compound.

Shape of ovary,—long, short, globular, flattened, compressed, lenticular.

Placentation,—axile, parietal, free-central.

Fruit,—berry, drupe, pome, akene, nut, caryopsis, follicle, legume, capsule, silique, silicle, cone, aggregate, collective.

HABITAT. Fields, marsh, woods, etc.

LOCALITY. Town and State.

STEM.

No. ..... Date..............................

Class ....Character ....

LEAF.

Insertion ....

Venation ....

Form ....

Margin ....

Upper surface ....

Lower surface ....

LEAF.

FLOWER.

Inflorescence .... Perfectness ....

Regularity .... Cohesion ....

No. of sepals .... No. of petals ....

Shape of corolla .... Color ....

Insertion of stamens ...... No. of stamens ....

OVARY.

No. of carpels ....

Cohesion ....

Shape of ovary ....

No. of cells .... No. of ovules ....

Placentation ....

Fruit ....

FLOWER.

IDENTIFICATION.

Family or Order ....

Genus .... Species ....

Common Name ....

Habitat .... Locality ....

Remarks ....

STEM.

No. ... Date....................

Class.................. .... Character. ....

LEAF.

Insertion

Venation .. ..

Form......... ....

Margin ....

Upper surface ....

Lower surface ....

LEAF.

FLOWER.

Inflorescence .... Perfectness ....

Regularity.............. .... Cohesion ....

No. of sepals.. No. of petals......................................

Shape of corolla .... Color..

Insertion of stamens.................................. No. of stamens....................................

OVARY.

No. of carpels ....

Cohesion ....

Shape of ovary...

No. of cells.................................No. of ovules...

Placentation...

Fruit...

FLOWER.

IDENTIFICATION.

Family or Order ....

Genus .... Species ....

Common Name ....

Habitat .... Locality ....

Remarks

STEM.

No. .... Date..................

Class ....Character ......

LEAF.

Insertion ....

Venation ....

Form ....

Margin ....

Upper surface ....

Lower surface ....

LEAF.

FLOWER.

Inflorescence .... Perfectness....................................

Regularity .... Cohesion ....

No. of sepals.................................... No. of petals....................................

Shape of corolla .... Color....................................

Insertion of stamens .... No. of stamens....................................

OVARY.

No. of carpels ....

Cohesion ....

Shape of ovary..

No. of cells.................................No. of ovules.................................

Placentation..

Fruit..

FLOWER.

IDENTIFICATION.

Family or Order ....

Genus .... Species ....

Common Name ....

Habitat .... Locality....................................

Remarks ....

STEM.

No. .... Date.........

Class... Character ....

LEAF.

Insertion ....

Venation ....

Form ....

Margin ....

Upper surface ....

Lower surface ....

LEAF.

FLOWER.

Inflorescence .... Perfectness ....

Regularity .... Cohesion ....

No. of sepals .... No. of petals ....

Shape of corolla .... Color ....

Insertion of stamens .... No. of stamens ....

OVARY.

No. of carpels ....

Cohesion ....

Shape of ovary ....

No. of cells .... No. of ovules ....

Placentation ....

Fruit ....

FLOWER.

IDENTIFICATION.

Family or Order ....

Genus .... Species ....

Common Name ....

Habitat .... Locality ....

Remarks ....

STEM.
No. ...... Date....
Class ....Character. ....

LEAF.
Insertion ...
Venation ....
Form ....
Margin ....
Upper surface ....
Lower surface ....

LEAF.

FLOWER.
Inflorescence .... Perfectness..
Regularity .... Cohesion ....
No. of sepals.. No. of petals ....
Shape of corolla .... Color..
Insertion of stamens.. No. of stamens ....

OVARY.
No. of carpels ....
Cohesion ....
Shape of ovary..
No. of cells..............................No. of ovules..............................
Placentation..
Fruit..

FLOWER.

IDENTIFICATION.
Family or Order ....
Genus .... Species ....
Common Name ....
Habitat .... Locality..
Remarks ....

STEM.

No. .... Date ....

Class .... Character ....

LEAF.

Insertion ....

Venation ....

Form ....

Margin ....

Upper surface ....

Lower surface ....

LEAF

FLOWER.

Inflorescence .... Perfectness ....

Regularity .... Cohesion ....

No. of sepals .... No. of petals ....

Shape of corolla .... Color ....

Insertion of stamens No. of stamens ....

OVARY.

No. of carpels ....

Cohesion ....

Shape of ovary ....

No. of cells .... No. of ovules ....

Placentation ....

Fruit ....

FLOWER.

IDENTIFICATION.

Family or Order ....

Genus .... Species ....

Common Name ....

Habitat .... Locality ....

Remarks ....

STEM.

No. .... Date ....

Class .... Character ....

LEAF.

Insertion ....

Venation ....

Form ....

Margin ....

Upper surface ....

Lower surface ....

LEAF.

FLOWER.

Inflorescence .... Perfectness ....

Regularity .... Cohesion ....

No. of sepals .... No. of petals ....

Shape of corolla .... Color ....

Insertion of stamens .... No. of stamens ....

OVARY.

No. of carpels ....

Cohesion ....

Shape of ovary ....

No. of cells .... No. of ovules ....

Placentation ....

Fruit ....

FLOWER.

IDENTIFICATION.

Family or Order ....

Genus .... Species ....

Common Name ....

Habitat .... Locality ....

Remarks ....

STEM.

No. .... Date........

Class .... Character ....

LEAF.

Insertion ....

Venation ....

Form ....

Margin ....

Upper surface ....

Lower surface ....

LEAF

FLOWER.

Inflorescence .... Perfectness ....

Regularity .... Cohesion ....

No. of sepals .... No. of petals ....

Shape of corolla .... Color ....

Insertion of stamens .... No. of stamens ....

OVARY.

No. of carpels ....

Cohesion ....

Shape of ovary ....

No. of cells .... No. of ovules ....

Placentation ....

Fruit ....

FLOWER.

IDENTIFICATION.

Family or Order ....

Genus .... Species ....

Common Name ....

Habitat .... Locality ....

Remarks ....

STEM.

No. .... Date ....

Class .... Character ....

LEAF.

Insertion ....

Venation ....

Form ....

Margin ....

Upper surface ....

Lower surface ....

LEAF.

FLOWER.

Inflorescence	Perfectness ....
Regularity	Cohesion ....
No. of sepals	No. of petals ....
Shape of corolla	Color ....
Insertion of stamens	No. of stamens ....

OVARY.

No. of carpels ....

Cohesion ....

Shape of ovary ....

No. of cells .... No. of ovules ....

Placentation ....

Fruit ....

FLOWER.

IDENTIFICATION.

Family or Order ....

Genus	Species ....

Common Name ....

Habitat	Locality ....

Remarks ....

STEM.

No. Date............

Class ......Character..........

LEAF.

Insertion ....

Venation ....

Form ....

Margin ....

Upper surface ....

Lower surface ....

LEAF.

FLOWER.

Inflorescence .... Perfectness ....

Regularity .... Cohesion ....

No. of sepals .... No. of petals ....

Shape of corolla .... Color ....

Insertion of stamens .... No. of stamens ....

OVARY.

No. of carpels ....

Cohesion ....

Shape of ovary ....

No. of cells ....No. of ovules ....

Placentation ....

Fruit ....

FLOWER.

IDENTIFICATION.

Family or Order ....

Genus .... Species ....

Common Name ....

Habitat .... Locality ....

Remarks ....

STEM.

No. .... Date ....

Class .... Character ....

LEAF.

Insertion ....

Venation ....

Form ....

Margin ....

Upper surface ....

Lower surface ....

LEAF.

FLOWER.

Inflorescence .... Perfectness ....

Regularity .... Cohesion ....

No. of sepals .... No. of petals ....

Shape of corolla .... Color ....

Insertion of stamens .... No. of stamens ....

OVARY.

No. of carpels ....

Cohesion ....

Shape of ovary ....

No. of cells .... No. of ovules ....

Placentation ....

Fruit ....

FLOWER.

IDENTIFICATION.

Family or Order ....

Genus .... Species ....

Common Name ....

Habitat .... Locality ....

Remarks ....

STEM.

No. .... Date.......

Class ....Character ....

LEAF.

Insertion ....

Venation ....

Form ....

Margin ....

Upper surface ....

Lower surface ....

LEAF.

FLOWER.

Inflorescence .... Perfectness..........

Regularity .... Cohesion ....

No. of sepals .... No. of petals ....

Shape of corolla .... Color ....

Insertion of stamens .... No. of stamens ....

OVARY.

No. of carpels ....

Cohesion ....

Shape of ovary ....

No. of cells.......... No. of ovules ....

Placentation ....

Fruit ....

FLOWER.

IDENTIFICATION.

Family or Order ....

Genus .... Species ....

Common Name ....

Habitat .... Locality ....

Remarks ....

STEM.
No. ... Date..........
Class...... Character ... .

LEAF.
Insertion .
Venation ... .
Form ... .
Margin ....
Upper surface ....
Lower surface ....

LEAF.

FLOWER.
Inflorescence .... Perfectness.................... ....
Regularity......... Cohesion ....
No. of sepals.......... No. of petals.......
Shape of corolla Color.......................... ....
Insertion of stamens . No. of stamens ....

OVARY.
No. of carpels ....
Cohesion ....
Shape of ovary..
No. of cells..................................No. of ovules..................................
Placentation...
Fruit..

FLOWER.

IDENTIFICATION.
Family or Order ....
Genus .... Species ....
Common Name
Habitat .... Locality...
Remarks ....

STEM.

No. .... Date ....

Class .... Character ....

LEAF.

Insertion ....

Venation ....

Form ....

Margin ....

Upper surface ....

Lower surface ....

LEAF.

FLOWER.

Inflorescence .... Perfectness ....

Regularity .... Cohesion ....

No. of sepals .... No. of petals ....

Shape of corolla .... Color ....

Insertion of stamens .... No. of stamens ....

OVARY.

No. of carpels ....

Cohesion ....

Shape of ovary ....

No. of cells .... No. of ovules ....

Placentation ....

Fruit ....

FLOWER.

IDENTIFICATION.

Family or Order ....

Genus .... Species ....

Common Name ....

Habitat .... Locality ....

Remarks ....

STEM.
No. ...... Date.................
Class .... Character ......

LEAF.
Insertion
Venation ....
Form ....
Margin ....
Upper surface ......
Lower surface ....

LEAF.

FLOWER.
Inflorescence .... Perfectness.. ....
Regularity.......... Cohesion ....
No. of sepals...... No. of petals
Shape of corolla .. Color..
Insertion of stamens.. No. of stamens ....

OVARY.
No. of carpels ....
Cohesion ....
Shape of ovary..
No. of cells.................................. No. of ovules....................................
Placentation..
Fruit..

FLOWER.

IDENTIFICATION.
Family or Order ....
Genus .... Species ....
Common Name ....
Habitat .... Locality
Remarks ....

STEM.

No. .... Date ....

Class .... Character ....

LEAF.

Insertion ....

Venation ....

Form ....

Margin ....

Upper surface ....

Lower surface ....

LEAF.

FLOWER.

Inflorescence .... Perfectness ....

Regularity .... Cohesion ....

No. of sepals .... No. of petals ....

Shape of corolla .... Color ....

Insertion of stamens .... No. of stamens ....

OVARY.

No. of carpels ....

Cohesion ....

Shape of ovary ....

No. of cells .... No. of ovules ....

Placentation ....

Fruit ....

FLOWER.

IDENTIFICATION.

Family or Order ....

Genus .... Species ....

Common Name ....

Habitat .... Locality ....

Remarks ....

STEM.

No. .... Date ....

Class .... Character ....

LEAF.

Insertion ....

Venation ....

Form ....

Margin ....

Upper surface ....

Lower surface ....

LEAF.

FLOWER.

Inflorescence .... Perfectness ....

Regularity .... Cohesion ....

No. of sepals .... No. of petals ....

Shape of corolla .... Color ....

Insertion of stamens .... No. of stamens ....

OVARY.

No. of carpels ....

Cohesion ....

Shape of ovary ....

No. of cells .... No. of ovules ....

Placentation ....

Fruit ....

FLOWER.

IDENTIFICATION.

Family or Order ....

Genus .... Species

Common Name ....

Habitat .... Locality ....

Remarks ....

STEM.

No. Date..........

Class .... Character ....

LEAF.

Insertion ....

Venation ....

Form ....

Margin ....

Upper surface ....

Lower surface ....

LEAF.

FLOWER.

Inflorescence .... Perfectness ....

Regularity .... Cohesion ....

No. of sepals .... No. of petals ....

Shape of corolla .... Color ....

Insertion of stamens .... No. of stamens ....

OVARY.

No. of carpels ....

Cohesion ....

Shape of ovary ....

No. of cells .... No. of ovules ....

Placentation ....

Fruit ....

FLOWER.

IDENTIFICATION.

Family or Order

Genus .... Species ....

Common Name ....

Habitat .... Locality ....

Remarks

STEM.

No. ...... Date ....

Class .... Character ....

LEAF.

Insertion ....

Venation ....

Form ....

Margin ....

Upper surface ....

Lower surface ....

LEAF.

FLOWER.

Inflorescence .... Perfectness ....

Regularity .... Cohesion ....

No. of sepals .... No. of petals ....

Shape of corolla .... Color ....

Insertion of stamens .... No. of stamens ....

OVARY.

No. of carpels ....

Cohesion ....

Shape of ovary ....

No. of cells .... No. of ovules ....

Placentation ....

Fruit ....

FLOWER.

IDENTIFICATION.

Family or Order ....

Genus .... Species ....

Common Name ....

Habitat .... Locality ....

Remarks ....

STEM.
No. Date
Class Character

LEAF.
Insertion
Venation
Form
Margin
Upper surface
Lower surface

LEAF.

FLOWER.
Inflorescence Perfectness
Regularity Cohesion
No. of sepals No. of petals
Shape of corolla Color
Insertion of stamens No. of stamens

OVARY.
No. of carpels
Cohesion
Shape of ovary
No. of cells No. of ovules
Placentation
Fruit

FLOWER.

IDENTIFICATION.
Family or Order
Genus Species
Common Name
Habitat Locality
Remarks

STEM.

No. Date...

Class ....Character...

LEAF.

Insertion

Venation ......

Form ....

Margin ....

Upper surface ...

Lower surface ....

LEAF.

FLOWER.

Inflorescence .... Perfectness...

Regularity .... Cohesion ....

No. of sepals.. No. of petals..

Shape of corolla .... Color..

Insertion of stamens ...... No. of stamens...

OVARY.

No. of carpels ....

Cohesion ....

Shape of ovary..

No. of cells.................................No. of ovules..................................

Placentation..

Fruit...

FLOWER.

IDENTIFICATION.

Family or Order ....

Genus .... Species ....

Common Name ....

Habitat .... Locality...

Remarks ....

STEM.
No. .... Date............

Class ....Character ....

LEAF.
Insertion ....

Venation ....

Form ....

Margin ....

Upper surface ....

Lower surface ....

LEAF

FLOWER.
Inflorescence .... Perfectness ....

Regularity .... Cohesion ....

No. of sepals .... No. of petals ....

Shape of corolla Color ....

Insertion of stamens .... No. of stamens ....

OVARY.
No. of carpels ....

Cohesion ....

Shape of ovary ....

No. of cells ....No. of ovules ....

Placentation ....

Fruit ....

FLOWER.

IDENTIFICATION.
Family or Order ....

Genus .... Species ....

Common Name ....

Habitat .... Locality ....

Remarks ....

STEM.
No. ...... Date.
Class ....Character

LEAF.
Insertion
Venation ....
Form. ....
Margin ....
Upper surface ...
Lower surface ....

LEAF.

FLOWER.
Inflorescence .... Perfectness..................
Regularity .... Cohesion ....
No. of sepals.................. No. of petals
Shape of corolla .... Color..................
Insertion of stamens ... No. of stamens

OVARY.
No. of carpels ....
Cohesion ....
Shape of ovary..................
No. of cells.................. No. of ovules..................
Placentation..................
Fruit..................

FLOWER.

IDENTIFICATION.
Family or Order ....
Genus .... Species ....
Common Name ....
Habitat .... Locality..................
Remarks ....

STEM.
No. .. Date..
Class ...... Character. ....

LEAF.
Insertion ....
Venation ....
Form ....
Margin ....
Upper surface ....
Lower surface ....

LEAF

FLOWER.
Inflorescence .... Perfectness ....
Regularity .... Cohesion ....
No. of sepals .... No. of petals ....
Shape of corolla Color ....
Insertion of stamens No. of stamens ....

OVARY.
No. of carpels ....
Cohesion ....
Shape of ovary ....
No. of cells .... No. of ovules ....
Placentation ....
Fruit ....

FLOWER.

IDENTIFICATION.
Family or Order ....
Genus .... Species ....
Common Name ....
Habitat .... Locality ....
Remarks ....

STEM.

No. .... Date

Class ...... Character ....

LEAF.

Insertion ....

Venation ....

Form ....

Margin ....

Upper surface ....

Lower surface ....

LEAF.

FLOWER.

Inflorescence .... Perfectness ....

Regularity .... Cohesion ....

No. of sepals .... No. of petals ....

Shape of corolla .... Color ....

Insertion of stamens .... No. of stamens ....

OVARY.

No. of carpels ....

Cohesion ....

Shape of ovary ....

No. of cells .... No. of ovules ....

Placentation ....

Fruit ....

FLOWER.

IDENTIFICATION.

Family or Order ....

Genus .... Species ....

Common Name ....

Habitat .... Locality ....

Remarks ....

STEM.

No. ...... Date........

Class ....Character ....

LEAF.

Insertion ....

Venation ....

Form ....

Margin ....

Upper surface ....

Lower surface ....

LEAF.

FLOWER.

Inflorescence .... Perfectness..........

Regularity .... Cohesion ....

No. of sepals.......... No. of petals..........

Shape of corolla .... Color..........

Insertion of stamens .... No. of stamens ....

OVARY.

No. of carpels ....

Cohesion ....

Shape of ovary ....

No. of cells.......... No. of ovules..........

Placentation..........

Fruit..........

FLOWER.

IDENTIFICATION.

Family or Order ....

Genus .... Species ....

Common Name ....

Habitat.......... Locality..........

Remarks ....

STEM.

No. Date

Class ......Character ...

LEAF.

Insertion ...

Venation ...

Form ...

Margin ...

Upper surface ...

Lower surface ...

LEAF.

FLOWER.

Inflorescence ... Perfectness ...

Regularity ... Cohesion ...

No. of sepals ... No. of petals ...

Shape of corolla ... Color ...

Insertion of stamens ... No. of stamens ...

OVARY.

No. of carpels ...

Cohesion ...

Shape of ovary ...

No. of cells ... No. of ovules ...

Placentation ...

Fruit ...

FLOWER.

IDENTIFICATION.

Family or Order ...

Genus ... Species ...

Common Name ...

Habitat ... Locality ...

Remarks ...

STEM.

No. .... Date......

Class ....Character ....

LEAF.

Insertion ....

Venation ....

Form ....

Margin ....

Upper surface ....

Lower surface ....

LEAF.

FLOWER.

Inflorescence .... Perfectness ....

Regularity .... Cohesion ....

No. of sepals .... No. of petals ....

Shape of corolla .... Color ....

Insertion of stamens .... No. of stamens ....

OVARY.

No. of carpels ....

Cohesion ....

Shape of ovary ....

No. of cells .... No. of ovules ....

Placentation ....

Fruit ....

FLOWER.

IDENTIFICATION.

Family or Order ....

Genus .... Species ....

Common Name ....

Habitat .... Locality ....

Remarks ....

STEM.

No. .... Date.......... ..

Class....... Character.

LEAF.

Insertion ...

Venation ....

Form .....

Margin ....

Upper surface

Lower surface ..

LEAF.

FLOWER.

Inflorescence .... Perfectness...

Regularity .... Cohesion ....

No. of sepals...... No. of petals......

Shape of corolla Color......................

Insertion of stamens. No. of stamens

OVARY.

No. of carpels ....

Cohesion ....

Shape of ovary......................................

No. of cells...... .... No. of ovules......................................

Placentation......................................

Fruit......................................

FLOWER.

IDENTIFICATION.

Family or Order ....

Genus .... Species ....

Common Name

Habitat .... Locality......................................

Remarks ....

STEM.

No. .... Date..

Class ....Character ....

LEAF.

Insertion ....

Venation ....

Form ....

Margin ....

Upper surface ....

Lower surface ....

LEAF.

FLOWER.

Inflorescence .... Perfectness..

Regularity.. Cohesion ....

No. of sepals.............. No. of petals....

Shape of corolla .... Color..

Insertion of stamens .... No. of stamens..

OVARY.

No. of carpels ....

Cohesion ....

Shape of ovary..

No. of cells.............................No. of ovules..............................

Placentation..

Fruit..

FLOWER.

IDENTIFICATION.

Family or Order

Genus .... Species ....

Common Name ....

Habitat .... Locality.

Remarks ....

STEM.

No. Date

Class Character

LEAF.

Insertion

Venation

Form

Margin

Upper surface

Lower surface

LEAF.

FLOWER.

Inflorescence Perfectness

Regularity Cohesion

No. of sepals No. of petals

Shape of corolla Color

Insertion of stamens No. of stamens

OVARY.

No. of carpels

Cohesion

Shape of ovary

No. of cells No. of ovules

Placentation

Fruit

FLOWER.

IDENTIFICATION.

Family or Order

Genus Species

Common Name

Habitat Locality

Remarks

STEM.

No. .... Date ....

Class .... Character ....

LEAF.

Insertion ....

Venation ....

Form ....

Margin ....

Upper surface ....

Lower surface ....

LEAF.

FLOWER.

Inflorescence .... Perfectness ....

Regularity .... Cohesion ....

No. of sepals .... No. of petals ....

Shape of corolla Color ....

Insertion of stamens .... No. of stamens ....

OVARY.

No. of carpels ....

Cohesion ....

Shape of ovary ....

No. of cells .... No. of ovules ....

Placentation ....

Fruit ....

FLOWER.

IDENTIFICATION.

Family or Order ....

Genus .... Species ....

Common Name ....

Habitat .... Locality

Remarks ....

STEM.

No. .... Date..

Class...Character..

LEAF.

Insertion ....

Venation ....

Form ....

Margin ....

Upper surface ....

Lower surface ....

LEAF.

FLOWER.

Inflorescence .... Perfectness....................................

Regularity.................................... Cohesion ....

No. of sepals.................................... No. of petals....................................

Shape of corolla .... Color....................................

Insertion of stamens ... No. of stamens....................................

OVARY.

No. of carpels ....

Cohesion ....

Shape of ovary..

No. of cells..............................No. of ovules..............................

Placentation..

Fruit..

FLOWER.

IDENTIFICATION.

Family or Order ....

Genus .... Species ....

Common Name ....

Habitat.................................... Locality....................................

Remarks ....

STEM.

No. .. Date.......... .

Class.... ....Character. .. ..

LEAF.

Insertion ....

Venation .

Form ....

Margin ....

Upper surface

Lower surface

LEAF.

FLOWER.

Inflorescence .... Perfectness

Regularity.......... Cohesion

No. of sepals... No. of petals

Shape of corolla Color

Insertion of stamens No. of stamens

OVARY.

No. of carpels

Cohesion

Shape of ovary

No. of cells............No. of ovules

Placentation

Fruit

FLOWER.

IDENTIFICATION.

Family or Order

Genus Species

Common Name

Habitat Locality

Remarks .

STEM.

No. .... Date....................

Class .... Character ....

LEAF.

Insertion ....

Venation ....

Form ....

Margin ....

Upper surface ....

Lower surface ....

LEAF.

FLOWER.

Inflorescence .... Perfectness ....

Regularity .... Cohesion ....

No. of sepals .... No. of petals ....

Shape of corolla .... Color ....

Insertion of stamens .... No. of stamens ....

OVARY.

No. of carpels ....

Cohesion ....

Shape of ovary ....

No. of cells .... No. of ovules ....

Placentation ....

Fruit ....

FLOWER.

IDENTIFICATION.

Family or Order ....

Genus .... Species ....

Common Name ....

Habitat .... Locality ....

Remarks ....

STEM.

No. ... Date ...

Class ... Character ...

LEAF.

Insertion ...

Venation ...

Form ...

Margin ...

Upper surface ...

Lower surface ...

LEAF.

FLOWER.

Inflorescence ... Perfectness ...

Regularity ... Cohesion ...

No. of sepals ... No. of petals ...

Shape of corolla ... Color ...

Insertion of stamens ... No. of stamens ...

OVARY.

No. of carpels ...

Cohesion ...

Shape of ovary ...

No. of cells ... No. of ovules ...

Placentation ...

Fruit ...

FLOWER.

IDENTIFICATION.

Family or Order ...

Genus ... Species ...

Common Name ...

Habitat ... Locality ...

Remarks

STEM.
No. .... Date....
Class .... Character ....

LEAF.
Insertion ....
Venation ....
Form ....
Margin ....
Upper surface ....
Lower surface ....

LEAF.

FLOWER.
Inflorescence .... Perfectness ....
Regularity .... Cohesion ....
No. of sepals .... No. of petals ....
Shape of corolla .... Color ....
Insertion of stamens .... No. of stamens ....

OVARY.
No. of carpels ....
Cohesion ....
Shape of ovary ....
No. of cells .... No. of ovules ....
Placentation ....
Fruit ....

FLOWER.

IDENTIFICATION.
Family or Order ....
Genus .... Species ....
Common Name ....
Habitat .... Locality ....
Remarks ....

STEM.

No. Date......

Class ....Character ...

LEAF.

Insertion ....

Venation ....

Form ....

Margin ....

Upper surface ....

Lower surface ....

LEAF

FLOWER.

Inflorescence .... Perfectness................................

Regularity .... Cohesion ....

No. of sepals................................ No. of petals................................

Shape of corolla Color................................

Insertion of stamens .... No. of stamens................................

OVARY.

No. of carpels ....

Cohesion ....

Shape of ovary..

No. of cells...................No. of ovules...................

Placentation..

Fruit..

FLOWER.

IDENTIFICATION.

Family or Order ....

Genus .... Species ....

Common Name ....

Habitat .... Locality................................

Remarks ....

STEM.
No. Date.
Class Character

LEAF.
Insertion
Venation
Form
Margin
Upper surface
Lower surface

LEAF.

FLOWER.
Inflorescence Perfectness
Regularity Cohesion
No. of sepals No. of petals
Shape of corolla Color
Insertion of stamens No. of stamens

OVARY.
No. of carpels
Cohesion
Shape of ovary
No. of cells No. of ovules
Placentation
Fruit

FLOWER.

IDENTIFICATION.
Family or Order
Genus Species
Common Name
Habitat Locality
Remarks

STEM.

No. .... Date ....

Class .... Character ....

LEAF.

Insertion ....

Venation ....

Form ....

Margin ....

Upper surface ....

Lower surface ....

LEAF

FLOWER.

Inflorescence .... Perfectness ....

Regularity .... Cohesion ....

No. of sepals .... No. of petals ....

Shape of corolla .... Color ....

Insertion of stamens .... No. of stamens ....

OVARY.

No. of carpels ....

Cohesion ....

Shape of ovary ....

No. of cells .... No. of ovules ....

Placentation ....

Fruit ....

FLOWER.

IDENTIFICATION.

Family or Order ....

Genus .... Species ....

Common Name ....

Habitat .... Locality ....

Remarks ....

STEM.

No. .... Date ....

Class .... Character ....

LEAF.

Insertion ....

Venation ....

Form ....

Margin ....

Upper surface ....

Lower surface ....

LEAF.

FLOWER.

Inflorescence .... Perfectness ....

Regularity .... Cohesion ....

No. of sepals .... No. of petals ....

Shape of corolla .... Color ....

Insertion of stamens .... No. of stamens ....

OVARY.

No. of carpels ....

Cohesion ....

Shape of ovary ....

No. of cells .... No. of ovules ....

Placentation ....

Fruit ....

FLOWER.

IDENTIFICATION.

Family or Order ....

Genus .... Species ....

Common Name ....

Habitat .... Locality ....

Remarks ....

STEM.

No. .... Date........

Class. .... Character........

LEAF.

Insertion ....

Venation ....

Form ....

Margin ....

Upper surface ....

Lower surface ....

LEAF.

FLOWER.

Inflorescence .... Perfectness ....

Regularity .... Cohesion ....

No. of sepals .... No. of petals ....

Shape of corolla .... Color ....

Insertion of stamens .... No. of stamens ....

OVARY.

No. of carpels ....

Cohesion ....

Shape of ovary ....

No. of cells .... No. of ovules ....

Placentation ....

Fruit ....

FLOWER.

IDENTIFICATION.

Family or Order ....

Genus .... Species ....

Common Name ....

Habitat .... Locality ....

Remarks ....

STEM.

No. .... Date

Class ....Character

LEAF.

Insertion ....

Venation ....

Form ....

Margin ....

Upper surface ....

Lower surface ....

LEAF.

FLOWER.

Inflorescence .... Perfectness ....

Regularity .... Cohesion ....

No. of sepals .... No. of petals ....

Shape of corolla .... Color ....

Insertion of stamens .... No. of stamens ....

OVARY.

No. of carpels ....

Cohesion ....

Shape of ovary ....

No. of cells .... No. of ovules ....

Placentation ....

Fruit ....

FLOWER.

IDENTIFICATION.

Family or Order ....

Genus .... Species ......

Common Name ....

Habitat .... Locality ....

Remarks ....

STEM.

No. .... Date........

Class .... Character ....

LEAF.

Insertion ....

Venation ....

Form ....

Margin ....

Upper surface ....

Lower surface ....

LEAF.

FLOWER.

Inflorescence .... Perfectness ....

Regularity .... Cohesion ....

No. of sepals .... No. of petals ....

Shape of corolla .... Color ....

Insertion of stamens .... No. of stamens ....

OVARY.

No. of carpels ....

Cohesion ....

Shape of ovary ....

No. of cells........................No. of ovules ....

Placentation ....

Fruit ....

FLOWER.

IDENTIFICATION.

Family or Order ....

Genus .... Species ....

Common Name ....

Habitat .... Locality ....

Remarks ....

STEM.

No. ... Date..........

Class.............................. .Character ....

LEAF.

Insertion ...

Venation ....

Form ... .

Margin ... .

Upper surface ....

Lower surface ....

LEAF.

FLOWER.

Inflorescence ..... Perfectness... ....

Regularity....... Cohesion ....

No. of sepals...... No. of petals.......

Shape of corolla Color.............................. ....

Insertion of stamens No. of stamens ....

OVARY.

No. of carpels ....

Cohesion ....

Shape of ovary..

No. of cells.............................No. of ovules..................................

Placentation..

Fruit..

FLOWER.

IDENTIFICATION.

Family or Order ....

Genus .... Species ....

Common Name

Habitat .... Locality..

Remarks ....

STEM.

No. .... Date..

Class . ...Character ....

LEAF.

Insertion

Venation ....

Form ....

Margin ....

Upper surface ....

Lower surface ....

LEAF.

FLOWER.

Inflorescence .... Perfectness......

Regularity.................... Cohesion ....

No. of sepals....... No. of petals....

Shape of corolla Color..............................

Insertion of stamens No. of stamens...........................

OVARY.

No. of carpels ....

Cohesion ....

Shape of ovary...

No. of cells.....................No. of ovules..................

Placentation...

Fruit..

FLOWER.

IDENTIFICATION.

Family or Order ....

Genus .... Species ....

Common Name ....

Habitat .... Locality.

Remarks ....

STEM.
No. Date
Class Character

LEAF.
Insertion
Venation
Form
Margin
Upper surface
Lower surface

LEAF.

FLOWER.
Inflorescence Perfectness
Regularity Cohesion
No. of sepals No. of petals
Shape of corolla Color
Insertion of stamens No. of stamens

OVARY.
No. of carpels
Cohesion
Shape of ovary
No. of cells No. of ovules
Placentation
Fruit

FLOWER.

IDENTIFICATION.
Family or Order
Genus Species
Common Name
Habitat Locality
Remarks

STEM.

No. .... Date........

Class .... Character ....

LEAF.

Insertion ....

Venation ....

Form ....

Margin ....

Upper surface ....

Lower surface ....

LEAF.

FLOWER.

Inflorescence .... Perfectness ....

Regularity .... Cohesion ....

No. of sepals .... No. of petals ....

Shape of corolla .... Color ....

Insertion of stamens .... No. of stamens ....

OVARY.

No. of carpels ....

Cohesion ....

Shape of ovary ....

No. of cells .... No. of ovules ....

Placentation ....

Fruit ....

FLOWER.

IDENTIFICATION.

Family or Order ....

Genus .... Species ....

Common Name ....

Habitat .... Locality ....

Remarks ....

STEM.

No. Date

Class Character

LEAF.

Insertion

Venation

Form

Margin

Upper surface

Lower surface

LEAF

FLOWER.

Inflorescence Perfectness

Regularity Cohesion

No. of sepals No. of petals

Shape of corolla Color

Insertion of stamens No. of stamens

OVARY.

No. of carpels

Cohesion

Shape of ovary

No. of cells No. of ovules

Placentation

Fruit

FLOWER.

IDENTIFICATION.

Family or Order

Genus Species

Common Name

Habitat Locality

Remarks

STEM.

No. Date.........

Class ....Character

LEAF.

Insertion

Venation

Form ....

Margin ....

Upper surface .....

Lower surface

LEAF

FLOWER.

Inflorescence Perfectness ....

Regularity Cohesion

No. of sepals No. of petals

Shape of corolla Color

Insertion of stamens No. of stamens.

OVARY.

No. of carpels ..

Cohesion

Shape of ovary

No. of cells. No. of ovules ..

Placentation

Fruit

FLOWER.

IDENTIFICATION.

Family or Order

Genus ... Species ..

Common Name

Habitat Locality.

Remarks

DRAWINGS.	DESCRIPTIONS.
(1)	
	2
	3
	4
	5
	6
	7
	8
	9
	10

DRAWINGS.	DESCRIPTIONS.
(1)	
	2
	3
	4
	5
	6
	7
	8
	9

DRAWINGS.	DESCRIPTIONS.
(1)	
	2
	3
	4
	5
	6
	7
	8
	9
	10

(1)	
	2
	3

2

3

4

5

6

7

8

9

10

2

3

4

5

6

7

8

9

DRAWINGS. DESCRIPTIONS.

(1)

2

3

4

5

6

7

8

9

10

(1)

2

3

4

5

6

7

8

DRAWINGS.	DESCRIPTIONS.
(1)	
	2
	3
	4
	5
	6
	7
	8
	9
	10

DRAWINGS.	DESCRIPTIONS.
(1)	
	2
	3
	4
	5
	6
	7
	8
	9

DRAWINGS.	DESCRIPTIONS.
(1)	
	2
	3
	4
	5
	6
	7
	8
	9
	10

DRAWINGS.	DESCRIPTIONS.
(1)	
	2
	3
	4
	5
	6
	7
	8
	9
	10

DRAWINGS.	DESCRIPTIONS.
(1)	
	2
	3
	4
	5
	6
	7
	8
	9
	10

(1)	
	2
	3
	4
	5
	6
	7

2

3

4

5

6

7

8

9

10

2

3

4

5

6

7

8

9

DRAWINGS.	DESCRIPTIONS.
(1)	
	2
	3
	4
	5
	6
	7
	8
	9
	10

DRAWINGS.	DESCRIPTIONS.
(1)	
	2
	3
	4
	5
	6
	7
	8
	9
	10

DRAWINGS.	DESCRIPTIONS.
(1)	
	2
	3
	4
	5
	6
	7
	8
	9
	10

(1)	
	2
	3
	4
	5
	6
	7
	8

DRAWINGS.	DESCRIPTIONS.
(1)	
	2
	3
	4
	5
	6
	7
	8
	9
	10

(1)	
	2
	3
	4
	5
	6
	7
	8
	9
	10

(1)

2

3

4

5

6

7

8

9

10

---

(1)

2

3

4

5

6

7

8

9

10

DRAWINGS.	DESCRIPTIONS.
(1)	
	2
	3
	4
	5
	6
	7
	8
	9
	10

(1)	
	2
	3

DRAWINGS.	DESCRIPTIONS.
(1)	
	2
	3
	4
	5
	6
	7
	8
	9
	10

DRAWINGS.	DESCRIPTIONS.
(1)	
	2
	3
	4
	5
	6
	7
	8
	9

DRAWINGS.	DESCRIPTIONS.
(1)	
	2
	3
	4
	5
	6
	7
	8
	9
	10

DRAWINGS.	DESCRIPTIONS.
(1)	
	2
	3
	4
	5
	6
	7
	8

DRAWINGS.	DESCRIPTIONS.
(1)	
	2
	3
	4
	5
	6
	7
	8
	9
	10

(1)	
	2
	3
	4
	5
	6
	7
	8

DRAWINGS. DESCRIPTIONS.

(1)

2

3

4

3

6

7

8

9

10

(1)

2

3

4

5

6

7

8

DRAWINGS.	DESCRIPTIONS.
(1)	
	2
	3
	4
	5
	6
	7
	8
	9
	10
(1)	
	2
	3
	4
	5
	6
	7
	8
	9
	10

2

3

4

5

6

7

8

9

10

(1)

2

3

4

5

6

7

8

DRAWINGS.	DESCRIPTIONS.
(1)	
	2
	3
	4
	5
	6
	7
	8
	9
	10

www.ingramcontent.com/pod-product-compliance
Lightning Source LLC
Chambersburg PA
CBHW022142090426
42742CB00010B/1358
* 9 7 8 3 3 3 7 7 1 1 9 4 8 *